Project IBIS

The Transresonator
Harmonic Frequency Induction
&
Transmagnetic Resonance Fields

By Timothy E. Douglas
Copyright 2024 by Timothy E. Douglas

While every precaution has been taken in the preparation of this book, the publisher assumes no responsibility for errors or omissions, or for damages resulting from the use of the information contained herein.

PROJECT I.B.I.S

First edition. August 5, 2024.

ISBN: 979-8227756664

Written by Timothy E. Douglas.

Part One:

<u>Original Thesis & Project Overview:</u>

Thesis:

This thesis investigates the hypothesis that ancient artifacts such as the "gods' handbags," the staff of Hermes, and the Ark of the Covenant were designed as Biomagnetic field manipulators. By examining the potential functions and historical contexts of these artifacts, and correlating them with modern scientific understanding of Biomagnetic fields and energy manipulation, this study aims to uncover how ancient civilizations might have used advanced knowledge of electromagnetism and biomagnetism. It is my belief that relics such as Tibetan Singing Bowls, The Bagdad Jar Battery – found in 1936, and Rotating Dial Mechanisms are the offsprings of the "God's Handbags" relic in particular.

Project:

This project aims to develop a modern interpretation of ancient artifacts, hypothesized to function as personal Biomagnetic field manipulators. The primary objective is to create a staff inspired by the Caduceus of Hermes, integrating sound frequency, electromagnetism, and Bio-resonance therapy principles. It is my belief that combining Sound Frequency, Electromagnetism, and Light Photons will not only produce anti-gravity effects, but also if tuned to the Schumann Wave Frequency (7.83hz) will reset the bodies Biomagnetic field to its original state. If tuned to the Solfeggio Frequency of 963hz – the frequency associated with the crown chakra and spiritual attunement- will enhance the effects and bring forth the true spiritual body that is within.

Thesis Explained

Introduction:

Esteemed reviewer, my name is Tim Douglas. Throughout my life I always thought that electromagnetism was the key to advancing our technology far beyond what we have been capable of. It wasn't until I started doing a delve into world religions and history dating back to Ancient Sumera and Egypt that the dots started to connect. During the course of my research and study into the ancient civilizations and spiritual realm, I discovered the missing parts to the primordial formula. To which I will discuss and show throughout this thesis.

The Earth's Schumann resonance and higher frequencies, such as 963 Hz, have been linked to various positive effects on human health and consciousness. The Schumann resonance aligns with the natural rhythms of the Earth, while 963 Hz is associated with spiritual awakening and mental clarity. Previous studies suggest that EM fields at these frequencies can influence biological systems, promoting relaxation, reducing stress, and enhancing overall well-being.

Reading and studying texts and scriptures such as the Kybalion, Emerald Tablets of Thoth, and the Gateway Experience document released by the CIA in 2010, gave me the understanding and clarity to start this research and development document. The goal of this thesis is to provide contextual understanding on how ancient technologies can be incorporated into modern day technologies providing breakthroughs in antigravity and Biomagnetic field manipulation for healing the physical and spiritual bodies.

The thesis will be broken down into two parts, Theoretical Background & Project Specifications. The project section will include plans and details that incorporate these theories and combine them into one device capable of Biomagnetic Field Manipulation for holistic healing. During my research, virtual experiments, and analysis' the success rate of these combined theories are 100% if done properly, however, it goes down to 82.3% if improper equipment and materials are used. My goal is to build and test a proof-of-concept device. Upon successful completion and documentation of the proof of concept, I will be looking for benefactors and highly capable individuals for the manufacturing of the prototype device.

Theoretical Background
In this section we explore the ancient technologies that I believe were used to manipulate Biomagnetic and electromagnetic fields. Throughout texts and scriptures these technologies have been depicted as symbolism for power and healing. Well, what if they actually were used to heal and generate power? What if the legends of old are and have always been true all along and we have just been unable to discover them until now?
As the texts and scriptures say, these "God's" were giants and we are but mere grasshoppers in their sight. So please keep this in mind as we discuss our first "technology" the God's Handbag.

In this picture and ones similar to it, a depiction of handbag and wrist watch device is present throughout the Mesoamerica timeline. If we go on the presumption that the Ark of the Covenant is some sort of power source device, I believe these handbags are correlated in theory to the Ark. The Ark was made of gold and specific dimensions stated in the Bible and was said to hold the 10 Commandments. Gold is a highly conductive metal, if we assume the 10 commandment tablets were made of iron ore and salt water were used in the construction of the ark, those are the ingredients to electricity at a very early start. As seen in the Bagdad Battery Jar and Nikola Tesla's research and development of electromagnetism.

The Battery Jar of Bagdad consists of copper, an iron bar, electrolyte solution, an asphalt stopper, and the vessel to hold it all together. These ingredients combined create electricity. Now we know the ancients had electrical power. If we look at the God's Handbag picture, the handbag itself, highly resembles the front side of a car battery. A square box with two "terminals" on either side and "wire" connecting them.

Now you're probably saying to yourself that I am stretching this theory pretty far and I would have to agree with you if I were in your position. Please bear with me a little longer and I will explain even further. Now if we assume that the Handbag is an earlier version of the Bagdad Battery, what would it be connected to? Well, I believe it is connected to the wrist watch device that in certain depictions a wire can be seen running from the handle to the side of the wrist device.

I believe that the wrist device is the controller to the system. It is my belief that the wrist device is a combination of a Tibetan Singing bowl and a Resonance Fan Dial. Where the wire connects to the wrist device, I believe there is a magnet attached so when the dial is spun manually, the magnetic force takes over and provides a perpetual source of motion, creating that electromagnetic field. So, what would be the purpose of such a device?

The purpose of these devices, in my opinion, is for Biomagnetic Field Manipulation. In the Sumerian Tablets of Creation and the Emerald Tablets of Thoth, it is stated that the formula for antigravity is Frequency + Electromagnetism. Well let's not forget Light Photons as well. It is also stated that this formula is the formula for life itself. Combing frequency, electromagnetism, and light photons will be the key to not only Antigravity and Holistic practices, but the future of space travel and matter conversion. Replicators and Tractor beams from Star Trek will be an actual possibility using this formula. If one simply has the resonance frequencies, healing and creation will be at the fingertips of the wielder.

I plan to focus on the Schumann frequency and the Solfeggio frequency of 963hz. The theory being that if the device is set to earth's natural magnetic field, it could potentially jump start our astral bodies or spiritual self like rebooting a computer. Then a run through of a 963hz will enhance and strengthen the connection to the divine frequency.

<u>Schumann, Solfeggio, and Sound Therapy</u>

In my delve into the world religions and holistic practices, I learned about the chakra system and how sound therapy and certain frequencies effect our physical bodies. Putting these methods into practice, I found the benefit and truth behind this practice. After a meditation using the Solfeggio frequencies, I felt more grounded, at ease, and more connected to myself and the universal spirit. Now I have to mention, I was not raised on holistic beliefs. I was raised on a very hard Catholic belief system. The do not question God's will or you're going to burn for all eternity type of Catholicism. Which is another topic all together and not the purpose of this thesis.

The Schumann Resonance can be described as:

"Schumann resonances are the principal background in the part of the electromagnetic spectrum from 3 Hz through 60 Hz and appear as distinct peaks at extremely low frequencies around 7.83 Hz (fundamental), 14.3, 20.8, 27.3, and 33.8 Hz. Schumann resonances occur because the space between the surface of the Earth and the conductive ionosphere acts as a closed, although variable-sized waveguide. The limited dimensions of the Earth cause this waveguide to act as a resonant cavity for electromagnetic waves in the extremely low frequency band. The cavity is naturally excited by electric currents in lightning." - Wikipedia

<u>Solfeggio Frequencies</u>

"The solfeggio frequencies are part of the olden six-tone scale believed to have incorporated sacred music, inclusive of the famous and beautiful Gregorian Chants.

The unique tones and chants are found to impart spiritual blessings when they are played harmoniously. Every solfeggio tone comprises frequencies necessary for balancing energy, keeping the spirit, mind, and body in a perfect form of harmony.

Solfeggio is the use of sol-fa syllables to note scale tones, that is, solmization. Solfeggio is also known as a singing exercise where syllables will be used other than using texts. There exist two types of solfeggios: the movable-do and the fixed-do.

There are six main solfeggio frequencies. They are;

The 396 Hz for liberating one from fear and guilt

The 417 Hz frequencies for facilitating change and undoing situations

The 528 Hz for miracles and transformations like DNA repair

The 639 Hz frequencies for relationships and reconnecting

The 741 Hz solfeggio frequencies for getting solutions and expressing themselves

The 852 Hz frequencies for returning one to a spiritual order

The 963 HZ solfeggio frequencies create room for oneness and unity.

Benefits of the Solfeggio Frequencies

There are several benefits linked with the solfeggio frequencies. They include:

Relieving pain and tension: The 174 HZ frequencies relieves a person from tension and pain. When one is listening to this type of music, they will observe their breath gradually slowing down, making them very relaxed as they drift off to quiet meditation. Healing frequencies will relax the muscles, thereby alleviating tension or pain.

Safety, energy, and survival are linked with the solfeggio frequencies at 285 Hz. It is a frequency linked to the root Muladhara or Chakra. It is the body's most fundamental and primal energy center at the base of a person's spine. The Chakra regulates all the energy relating to safety, survival, and instinct.

The 396 Hz Solfeggio Frequencies release fear and guilt. When listening to the frequency, the feelings of fear and guilt kick in. A mantra can be used for harmony to be obtained from the ancient Solfeggio frequencies.

The 417 Hz releases negativity and all past trauma. One will experience relief because the negative experiences and influences from the past slowly leave their consciousness. It is linked to solar plexus chakra and color yellow.

528 Hz is for clarity, peace, and DNA healing: it is a miracle tone or the love frequencies which have been claimed to heal the DNA and also cleanses an individual from any diseases and sickness. Healing impact is amplified through toning and meditation.

Healing interpersonal relationships using 639 Hz: it is a frequency with an ability to reconnect because it allows one to meditate on themselves, their life, and those of the people they value and love.

Problem-solving and improving emotional stability using 741 Hz: it allows the mind to expand, filling it with various solutions and new ways for self-expression. It unwinds a person's worries and troubles, allowing one to think freely. It creates room for a healthier and stable emotional position and time for more profound intuition.

To create harmony with the universe and yourself: use the 852 HZ solfeggio frequencies so that you can heighten your intuitions and alignment with spiritual orders. You can connect with an omnipresent spirit in the universes and bring back harmony to yourself.

The 963 HZ solfeggio frequencies create room for oneness and unity. It awakens your interactions with oneness and interconnectedness with the universes. You get pure and bright visions and thoughts in your mind."

This explanation of the Solfeggio Frequencies comes from www.naturehealingsociety.com[1].

<u>Sound Therapy & Hemi-Sync</u>

Sound Therapy is the use of Tibetan Singing bowls, Tuning Forks, and the power of your own voice to create resonance frequencies that heal and rebalance your chakra system and physical body. Some Sound Therapists have created their own music and mediations using the Solfeggio Frequencies as a baseline then incorporates instruments or sounds of nature to produce a calming and relaxing meditation experience.

In 2010 the CIA released a document titled "The Gateway Experience" conducted by LTC Wayne M. McDonnell and further developed by Robert Monroe. This experiment was conducted in secret to research the effects of Sound Frequencies and Bi-neural Beats on human physiology. In this document it states the truth of this experiment and that sound frequencies do in fact alter or change our states of consciousness. The Double Split Experiment verifies that at the subatomic (electron) level, there is consciousness to all things. The fact that everything is in a state of vibration further enforces these notions.

---

1. http://www.naturehealingsociety.com

<u>Selections from the Gateway Experience Document</u>
Biofeedback:

is somewhat unique in that it actually employs the self-cognitive powers of the left hemisphere to gain access to such areas of the right brain as the lower cerebral, motor and sensory cortices and assorted pain or pleasure centers. Instead of suppressing the left hemisphere as is done in hypnosis, or largely bypassing and ignoring it as is done in transcendental meditation, biofeedback teaches the left hemisphere first to visualize the desired result and then to recognize the feelings associated with the experience of successful right hemisphere access to the specific lower cerebral, cortex, pain or pleasure or other areas in the manner needed to produce the desired result, Special self-monitoring devices such as the digital thermometer is used to inform the left brain when it succeeds in keying the right hemisphere into produce the same external, objective measures of success. In this way, the pathways are strengthened and emphasized to such an extent that left brain consciousness is enabled to access appropriate areas in the right brain using a the conscious, demand node. For example, if the subject wishes to increase the circulation in the left log in order to speed up healing he may concentrate with his left brain on achieving that result while carefully monitoring a digital thermometer connected to the left leg. When the concentrated effort begins to affirmation and repetition. In this way, pain can be blocked, healing car can be suppressed and ultimately destroyed, enhanced, malignant tumors can be the body's pleasure centers can be and a variety of specific physiological results may be achieved. In addition, biofeedback may be used to greatly accelerate achievement of deep meditative states particularly for beginners who have no experience in meditative techniques and whose progress in that methodology is enhanced through effective visualization and external, objective affirmation. Display of the subject's brainwave pattern on a cathode ray tube has proven to be a laboratory-validated means by which subjects may quickly learn to place themselves in profoundly relaxed states characterized by the sort of quietude and singularity of mental focus associated with advanced meditation.

Gateway and Hemi-Sync.

Now that we have briefly profiled the basic mechanics of the principal techniques for altering or expanding consciousness which share some of the objectives and/or methods employed in the Gateway Experience, we may proceed to focus on what that technique actually involves. Fundamentally, the Gateway Experience is training system designed to bring enhanced strength, focus and coherence to the amplitude and frequency of brainwave output between the left and right hemispheres so as to alter consciousness, moving it outside the physical sphere so as to ultimately escape even the restrictions of time and space. The participant then gains access to the various levels of intuitive knowledge which the universe offers. What differentiates the Gateway Experience from other forms of meditation, is its use of the Hemi-Sync technique which is defined in a monograph by Monroe Institute trainer Melissa Jager as, "a state of consciousness defined when the EEG patterns of both hemispheres are simultaneously equal in amplitude and frequency. Although Hemi-Sync seems to be rather rare and of only short duration in ordinary human consciousness, Melissa Jager states that: "Audio techniques developed by Bob Monroe can induce and sustain Hemi-Sync with the Institute's basic Focus 3 tapes...." she also notes that: "Studies conducted by Elmer and Alyce Greene at the Menninger Foundation have shown that a subject with 20 years of training in Zen meditation could consistently establish Hemi-Sync at will, sustaining it for over 15 minutes." Dr. Stuart Twemlow, a psychiatrist and research associate of the Monroe Institute, reports that: "In our studies of the effect of the Monroe tape system on brainwaves, we have found that the tapes encourage the focusing of brain energy (it can be measured as with a lightbulb, in watts) into narrower and narrower frequency bands."

This focusing of energy is not unlike the yoga concept of one pointedness, which we may translate in western terms as single-mindedness." Dr. Twemlow goes on to observe that as the individual gets into the tapes beyond Focus 3, "...there is gradual increase in brainwave size which is a measure of brain energy or power."

Frequency Following Response.

To achieve synchronization of brain hemispheres, the Hemi-Sync technique takes advantage of a phenomenon known as the Frequency Following Response (FFR) which means that if a subject hears a sound produced at a frequency which emulates one of those associated with the operation of the human brain, the brain will try to mimic the same frequency pattern by adjusting its brainwave output. Therefore, if the subject is in a fully awake state but hears sound frequencies which approximates brainwave output at the Theta level, the subject's brain will endeavor to alter its brainwave pattern from the normal Beta to the Theta level. since the Theta level is associated with sleep, the subject concerned may progress from fully awake to sleep state (provided that he does not consciously resist) as the brain strives to entrain its wave frequency output with the one which the person hears. Since these brainwave frequencies are outside the spectrum of sounds which can be heard in pure form by the human ear, Hemi-Sync must produce them based on another phenomenon known as the brain's capacity for deducing "beat" frequencies. If the human brain is exposed to one frequency in the left ear which is 20 Hertz below another audible frequency played in the right ear, rather than hearing either of the two audible frequencies, the brain chooses to "hear" the difference between them, the "beat" frequency. Thus, availing itself of the FFR phenomenon, and using the technique of "beat" frequencies, the Gateway system uses Hemi-Sync and other audio techniques employing the FFR phenomenon to introduce a variety of frequencies which are played at a virtually subliminal, marginally audible level. The objective is to relax the left hemisphere of the brain, place the physical body in a virtual sleep state, and bring the left and right hemispheres into coherence under conditions designed to promote the production of ever higher amplitude and frequency of brainwave output. Audible and perhaps subliminal suggestions by Bob Monroe accompany the various brainwave frequencies, which are sometimes rolled in together with other sounds such as sea surf to mask the sound frequencies where desirable. In this way, Gateway endeavors to provide the subject with the tools by which he may alter his consciousness based on his own volition over time through the repetitive use of the tapes so as to access, via intuitive means, new categories of information not available to ordinary consciousness.

Role of Resonance.

However, brain coherence through entrainment to "beat" frequencies introduced via stereo headphones is only part of the reason why the Gateway system works. It is also designed to achieve the physical quietude characteristic of deep transcendental meditative states which brings about a complete alteration of the fundamental resonance pattern associated with the sound frequencies produced by the human body. Yoga, Zen of transcendental meditation, if practiced long enough, will produce a change in the sound frequency with which the human heart resonates throughout the entire body. According to Bentov, this change in resonance results from elimination of what the medical profession calls "the bifurcation echo" so that the sound of the heartbeat can move synchronously up and down the circulatory system in harmonious resonance approximately seven time second. Bentov describes the roll played by the bifurcation echo as follows: "When the left ventricle of the heart ejects blood, the aorta, being elastic, balloons out just beyond the valve and causes a pressure pulse to travel down along the aorta. When the pressure pulse reaches the bifurcation in the lower abdomen (which is where the aorta forks in two to go into the legs), part of the pressure pulse rebounds and starts traveling up the aorta. If in the meantime the heart ejects more blood, and new pressure pulse is traveling down, these two pressure points will eventually collide somewhere along the aorta and produce an Interference pattern. By placing the body in a sleeplike state, the Gateway tapes achieve the same goal as meditation in that it places the body in such a profoundly relaxed state that the bifurcation echo slowly fades away as the heart lessens the force and frequency with which it pushes blood into the aorta. The result is a regular, rhythmic sinewave pattern of sound which echoes throughout the body and rises up into the head in sustained resonance. The amplitude of this sinewave pattern, when measured with a sensitive, seismograph type instrument is about three times the average of the sound volume produced by the heart when it is operating normally.

The Consciousness Matrix.
The universe is composed of interacting energy fields, some at rest and some in motion. It is, in and of itself, one gigantic hologram of unbelievable complexity. According to the theories of Karl Pribram, neuroscientist University of Stanford and David Bohm, a physicist at the University of London, the human mind is also a hologram which attunes itself to the universal hologram by the medium of energy exchange thereby deducing meaning and achieving the state which we call consciousness. With respect to states of expanded or altered consciousness such as Gateway uses, the process operates in the following way. As energy passes through various aspects of the universal hologram and is perceived by the electrostatic fields which comprise the human mind, the holographic images being conveyed are projected upon those electrostatic fields of the mind and are perceived or understood to the extent that the electrostatic field is operating at a frequency and amplitude that can harmonize with and therefore "read" the energy carrier wave pattern passing through it. Changes in the frequency and amplitude of the electrostatic field which comprises the human mind determines the configuration and hence the character of the holographic energy matrix which the mind projects to intercept meaning directly from the holographic transmissions of the universe. Then, to make sense of what the holographic image is "saying" to it, the mind proceeds to compare the image just received with itself. Specifically, it does this by comparing the image received with that part of its own hologram which constitutes memory. By registering differences in geometric form and in energy frequency, the consciousness perceives. As psychologist Keith Floyd puts it "Contrary to what everyone knows in so, it may not be the brain that produces consciousness—but rather, consciousness that creates the appearance of the brain..."
The Gateway Experience Document from the CIA further proves and verifies the validity of Sound Therapy and the effects frequency has on

our bodies and mind. Now that we have the ancient technology and sciences behind this theory, I will now move into the Project Phase of this thesis.

<u>Project IBIS: Device Overview</u>
<u>Proof of Concept:</u>
<u>Basic Materials for Proof of Concept</u>
Bluetooth Speaker:
A small, portable Bluetooth speaker or a wired speaker that can connect to your phone.
Phone as Tone Generator:
A smartphone with a tone generator app (there are many free apps available for both Android and iOS).
200n Electromagnet or Tesla Coil:
Can be found and bought on Amazon
Power Source:
A battery pack or a power supply suitable for your Tesla coil or electromagnet
Non-Metallic Tube or Rod:
A plastic or wooden dowel (around 1-2 feet long) to serve as the base structure for mounting your components.
<u>Assembly and Testing Steps</u>
Setting Up the Speaker:
Connect the speaker to your phone via Bluetooth or a wired connection.
Use the tone generator app on your phone to produce specific frequencies, such as the Schumann resonance 7.83 Hz or 963 Hz.
Combining Components:
Position the speaker, electromagnet or Tesla coil together on the non-metallic tube or rod, securing them with tape or zip ties.
Make sure the components are positioned so that the sound from the speaker can influence the electromagnetic coil and vice versa.
Conducting the Test:
Power on the Tesla coil to generate the electromagnetic field.
Use the tone generator app on your phone to play specific frequencies through the speaker.

Observe the interactions between the sound frequencies and the electromagnetic field. You might notice changes in sound quality, vibrations, or electromagnetic effects.

Observations and Adjustments:

Sound and Vibration: Listen for any changes in sound quality or vibrations when the electromagnetic field is active.

Electromagnetic Field: Use a simple EMF detector app on another phone or an EMF meter to observe the electromagnetic field's strength and behavior.

Adjustments:

Experiment with different frequencies, coil configurations, and power levels to observe various effects.

Safety Considerations:

Tesla Coil Safety: Tesla coils can generate high voltage, so handle them with care and follow all safety guidelines provided with the kit.

Power Source: Ensure your power source is appropriate for the Tesla coil and does not exceed recommended voltage and current levels.

<u>Conclusion</u>

This small-scale test setup using basic materials can help you understand the interactions between sound frequencies, resonance, and electromagnetic fields. It provides a tangible proof of concept before moving on to a more complex and fully realized prototype.

Once the concept is verified, further research and development for a prototype device will need to commence.

<u>Potential Effects of Biomagnetic Field Manipulation</u>
Resonance Characteristics:
At 963 Hz, the speaker will produce a higher-pitched sound compared to the Schumann resonance.
The electromagnetic field generated by the copper wiring at this frequency may exhibit different interaction dynamics with the magnetite and the surrounding environment.
Field Intensity and Distribution:
The higher frequency might result in a more localized and intense electromagnetic field around the vessel.
This could influence the spatial distribution of the field, potentially creating areas of higher energy concentration.
Energy Dynamics:
The energy dynamics at 963 Hz may differ from those at lower frequencies, potentially affecting the overall stability and coherence of the generated electromagnetic field.
The field might exhibit more rapid oscillations, which could influence how it interacts with biological tissues.

<u>Potential Effects on the Human Body</u>
Spiritual and Psychological Impact:
The 963 Hz frequency is often associated with awakening intuition and activating the pineal gland, which is linked to higher states of consciousness and spiritual awakening.
Exposure to this frequency might promote feelings of peace, connection to the divine, and heightened states of awareness.
Biofield Alignment:
The human body's biofield might resonate differently at 963 Hz compared to the Schumann resonance. This frequency could stimulate specific energy centers (chakras), particularly the crown chakra, which is associated with spiritual connection.

It could help in balancing and aligning the biofield in a way that promotes spiritual growth and mental clarity.

Physical and Emotional Well-Being:

Some studies and anecdotal evidence suggest that exposure to higher Solfeggio frequencies like 963 Hz can aid in reducing stress, enhancing mood, and improving overall well-being.

The specific effects can vary among individuals, but many report a sense of mental clarity, emotional release, and heightened intuition.

Ibis Staff Prototype Configuration

Outer Shell:

Material: Carbon Fiber or Reinforced Plastic

Reason: These materials are lightweight, durable, and strong, making the staff easy to handle and resilient to wear and tear.

Internal Frame:

Material: Aluminum or Titanium

Reason: Both materials are lightweight yet strong, providing structural integrity without adding significant weight.

Resonance Mechanism:

Material: High-Quality Metal Alloy (such as Bronze or Brass)

Reason: These metals have excellent acoustic properties, producing clear and resonant tones when vibrated.

Electromagnetic Coils

Material: Copper

Reason: Copper is an excellent conductor of electricity, making it ideal for generating electromagnetic fields.

Power Source

Type: Rechargeable Lithium-Ion Battery

Reason: These batteries have a high energy density, are lightweight, and provide reliable power.

Speakers

Type: High-Quality Piezoelectric or Electrodynamic Speakers

Reason: These speakers can produce clear and accurate sound at various frequencies.

<u>Internal Dimensions and Layout</u>

Overall Dimensions:

Length: 4 feet (48 inches)

Diameter: 2 inches

<u>Segmented Design:</u>

<u>Top Section (12 inches)</u>

Houses control interface (digital panel or Bluetooth module) and rechargeable battery pack.

Accessible via a removable cap for battery replacement and control adjustments.

<u>Middle Section (24 inches)</u>

Contains the resonance mechanism (resonance fan/dial) and sound production components (speakers).

Includes mounting points for the resonance mechanism and acoustic chambers for sound amplification.

<u>Bottom Section (12 inches)</u>

Encases the electromagnetic coils and any additional circuitry for generating the electromagnetic field.

Designed with ventilation and heat dissipation features to prevent overheating.

<u>Resonance Mechanism</u>

Diameter of Resonance Fan/Dial: 1.5 inches

Length of Resonance Mechanism Housing: 10 inches (within the middle section)

Materials: High-Quality Metal Alloy with precision-engineered blades or surfaces for optimal vibration and sound production.

The Resonance Dial spins as the Sound and Electromagnetic waves pass through it.

<u>Speakers</u>

Number: 2-4 small, evenly spaced within the middle section

Diameter of Each Speaker: 1-1.5 inches
Position: Mounted around the resonance mechanism to ensure even sound distribution.

<u>Electromagnetic Coils</u>
Length: 10 inches (within the bottom section)
Number of Coils: 3-6 concentric coils
Wire Gauge: 22 AWG copper wire
Turns Per Coil: 222 turns

<u>Assembly and Integration</u>
Modular Design:
Each section (top, middle, bottom) should be modular and detachable for easy assembly, maintenance, and upgrades.
Vibration Isolation:
Use rubber or silicone dampers to isolate the resonance mechanism from the outer shell to prevent unwanted vibrations and ensure clear sound production.
Heat Management:
Include heat sinks and ventilation holes in the bottom section to manage heat generated by the electromagnetic coils and power source.
Control Interface:
A digital control panel or Bluetooth module should be mounted in the top section for easy access.
Include settings for adjusting frequencies, power levels, and modes (e.g., Schumann resonance, 963 Hz, etc.).

<u>Final Considerations</u>
Ergonomics: Ensure the staff is comfortable to hold and use, with a well-balanced weight distribution.
Aesthetics: Consider incorporating decorative elements inspired by the caduceus, such as engravings or symbolic motifs.
Testing and Calibration: Conduct extensive testing and calibration to ensure the staff produces the desired frequencies and electromagnetic fields effectively.

<u>Conclusion</u>

By carefully selecting materials and designing the internal dimensions and layout, we can create a resonance staff that successfully integrates the principles of Tibetan singing bowls, Schumann resonance, the Solfeggio frequencies and electromagnetism. This innovative tool will offer a modern, portable solution for meditation, healing, and energy alignment.

<u>Potential Outcomes</u>

Resonance with Earth's Electromagnetic Field:

If the Staff is tuned to the Schumann resonance, it could create a harmonious interaction between its electromagnetic field and the Earth's natural frequencies.

This alignment might help realign or strengthen the human body's electromagnetic field, which naturally resonates at similar frequencies.

Biological Impact:

The human body responds to electromagnetic fields. Exposure to frequencies in the Schumann resonance range is believed to promote relaxation, reduce stress, and improve overall well-being.

If the experiment successfully produces a stable Schumann resonance, it could potentially enhance these positive effects when applied to the body.

Energy Realignment:

The electromagnetic field produced by the vessel, if tuned correctly, might help realign the body's biofield (the electromagnetic field generated by living organisms) with the Earth's natural frequencies. This realignment could promote balance and coherence in the body's energy systems, possibly leading to improved health and vitality.

Integration with Other Technologies:

The staff could potentially integrate with other digital health and wellness technologies, such as biofeedback devices or meditation apps. Data collected from these integrations could help refine and personalize the resonance effects for individual users.

Potential Therapeutic Applications:
The setup could be used in therapeutic settings to help individuals attune their electromagnetic fields to the Earth's resonance or Solfeggio frequencies.
Regular exposure to these frequencies might support holistic health practices; aiding in meditation, stress reduction, and energy healing.

<u>Potential Breakthrough and Technological Advancement</u>
Studying and researching the effects of frequency and electromagnetism doesn't just stop at holistic healing practices. It is my belief that the study and development of this technology will be the key to space travel or Star Gates, matter conversion, and teleportation. It's all about finding the right resonate frequency and having a device capable of manipulating bio-electrical-magnetic fields. Combining frequency, electromagnetism and light photons is the key to make these technologies go from Science Fiction to Science Fact.
<u>Practical Considerations</u>
Frequency Accuracy:
Ensure the Staff's resonant frequency is precisely tuned to the Schumann resonance as its base frequency. This may require fine-tuning the design and the applied electrical current.
Safety Measures:
Monitor the intensity of the electromagnetic field to avoid any adverse effects from overexposure. The field strength should be within safe limits for human exposure. Between 10n – 200n.

Measurement and Monitoring:
Use biofield sensors to measure changes in the human body's electromagnetic field before, during, and after exposure.
Track physiological and psychological indicators to assess the overall impact on well-being.

<u>Conclusion</u>
If my theory and experiment is successful, it could provide a novel way to enhance the alignment of the human body's electromagnetic field with the Earth's natural frequencies. This alignment might promote

various health benefits, making it a valuable tool in energy healing and holistic health practices.

<u>References</u>

The Gateway Experience, CIA, Wayne McDonnell & Robert Monroe

Emerald Tablets of Thoth a Literal Translation by Maurice Doreal

Sumerian Tablets of Creation

Sacred-Texts.com

Wheels of Life by Anodea Judith PhD

InternetArchives.org

Erich von Däniken's Chariots of the Gods

Magnetochemistry Journal on Recent Advancements in Biomagnetism
https://link.springer.com/chapter/10.1007/978-94-017-0373-4_2

HeartMath Institute Research on Bioelectromagnetic Communication
https://www.heartmath.org/research/research-library/energetics/energetic-heart-bioelectromagnetic-communication-within-and-between-people/

Reference Pictures

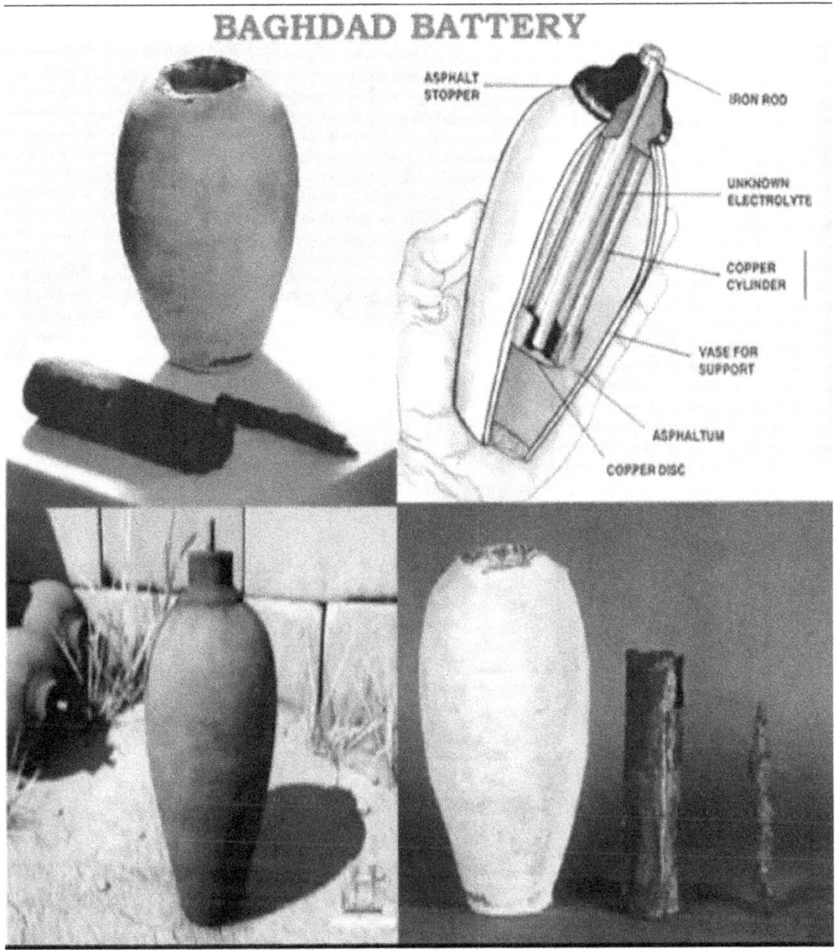

The Handbags of the Gods

Ark of the Covenant

Tibetan Singing Bowl

Caduceus Staff / Staff of Hermes

Sumerian Tablets

Emerald Tablet of Thoth the Atlantean

Emerald Tablets of Thoth - Lecturum

The Gateway Experience – The Absolute in Infinity & Consciousness Energy Grid

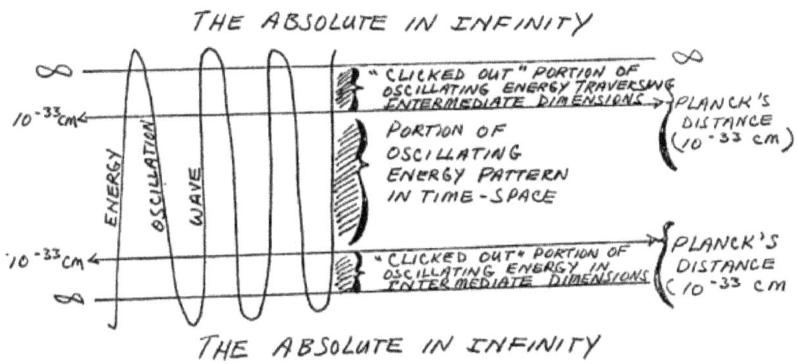

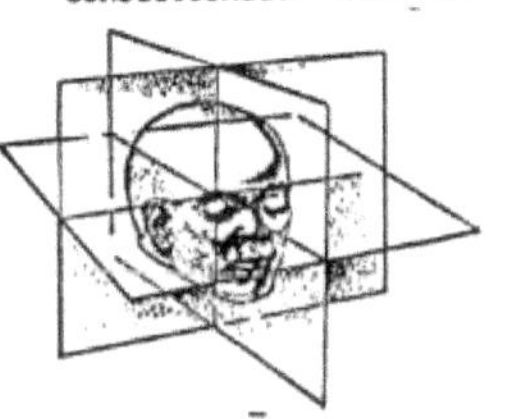

Left Hemisphere Consciousness Grid

Acts like the Mind's computer software to reduce input from right hemisphere to verbal symbols and concepts.

Right Hemisphere Consciousness Grid

Reduces three dimension holographic image to tw dimensional go/no go fo

Part Two:

<u>The Transresonator, Harmonic Frequency Induction, and Transmagnetic Resonance Fields</u>

<u>The Transresonator Device & Construction</u>
<u>Transresonator Construction:</u>

The transresonator device developed for this research is a prototype designed to test the hypothesis that specific resonant frequencies can enhance electromagnetic fields. The device was constructed using a combination of a Bluetooth speaker, an electromagnet, and additional magnet components to form the magnet stack. Below is a breakdown of the core elements and their role in the device:

1. Bluetooth Speaker:

The Bluetooth speaker serves as the base for generating and delivering frequency signals to the transresonator. It was repurposed for its connection to an external signal generator and amplifier setup, allowing the speaker's leads to drive the electromagnet directly.

Amplifier Leads: The internal wiring of the speaker was modified so that the amplifier leads could be connected directly to the electromagnet, bypassing the speaker cone. This allowed the speaker to act as a frequency delivery system without producing sound, purely transmitting the frequency signal into the electromagnetic setup. This method of inducting frequencies I have termed Harmonic Frequency Induction to produce Transmagnetic Resonance Fields.

2. 9V 50N Electromagnet:

A key component of the transresonator, the 9V 50N electromagnet was installed in place of the speaker's diaphragm. The electromagnet was essential for propagating the signal through the electromagnetic field (EMF) that could be influenced by the applied frequencies.

EM Field Properties: The electromagnet- combined with the speaker cone magnet, connected by a steel bolt, produced an initial field with a strength of approximately 0.42uT (microteslas) when no frequency was applied. Once the 144,000 Hz frequency was introduced, the field strength increased to 0.81uT, demonstrating the influence of frequency on EMF enhancement.

3. Internal Speaker Magnet:

It was observed that a stable magnetic field is needed for the propagation of the signal. Thus, then the internal magnet of the Bluetooth speaker remained, providing a consistent magnetic field that interacted with the newly connected electromagnet and frequency generator. This field, combined with the input from the signal generator, played a role in creating the resonance needed for testing.

4. Magnet Stack:

The magnet stack was added as part of the latest experimental iteration. This stack includes multiple magnets layered to increase the baseline EMF, providing a stronger field to test frequency effects on.

Baseline Field Strength: The magnet stack produced a base EMF of approximately 4uT, with peaks reaching as high as 18.64uT before any frequencies were applied.

Testing without Power or Signal: When no signal or power was applied, the magnet stack maintained this baseline field, which provided a stable foundation for comparing the effects of frequency input.

5. Signal Generator:

A critical aspect of the experiment involved the use of a signal generator capable of outputting various frequencies, including the solfeggio-based frequencies like 144,000 Hz.

Frequency Range: The generator used in these tests delivered signals within a specified range of harmonic frequencies, focusing on those theorized to be within the 3/6/9 resonance matrix.

Future Testing with Higher Frequencies:

Upcoming tests will involve a spectrum analyzer with a DDS generator capable of generating frequencies up to 350 MHz, allowing for more comprehensive exploration of high-frequency resonance impacts on the electromagnetic fields.

6. EMF Reader:

Throughout the experiment, an EMF reader was used to capture baseline readings and monitor changes in the magnetic field strength when frequencies were applied.

Results:

The EMF readings consistently showed an increase in field strength when resonant frequencies were introduced. For instance, the application of 144,000 Hz caused the field to nearly double in strength, from 0.42uT to 0.81uT. The 3 in 1 device used as a DDS generator, provided Oscilloscope and multi-meter readings as well. The wave pattern on the Oscilloscope matched that of the one sent through the speaker system at 20448Hz.

The field created from my method of Harmonic Frequency Induction is termed a Transmagnetic Resonance Field.

Note: Setup did not interfere with Bluetooth signal of speaker. All Frequencies under 22,000Hz were ran through the speaker from my cell phone via bluetooth.

7. Potential Modifications and Future Development:

Spectrum Analyzer: The next phase of testing will include a spectrum analyzer capable of measuring frequencies up to 350 MHz. This will enable exploration of higher-frequency effects on the magnetic field and allow for more precision in identifying resonant frequency thresholds.

Refining the Transresonator Design:

Future iterations of the transresonator will incorporate additional components, such as advanced signal processors and different magnet configurations, to optimize resonance efficiency and magnetic field strength.

Methodology & Applications of the Transresonator

Main Application: Resonance Frequency Alignment or Transmagnetic Resonance Therapy

The core function of the transresonator lies in its ability to alter an individual's resonance frequency, affecting both physical and energetic levels. Through the combination of precise electromagnetic fields and sound frequencies, the system induces changes on a subatomic level, causing the user's body and energetic field to resonate at the same frequency as that which is being output by the device. This matching of resonance frequencies could theoretically:

Enhance energetic healing by harmonizing with the body's natural frequencies.

Alter or raise the user's vibrational state, which could have implications for mental clarity, physical wellness, and spiritual ascension.

Serve as a mechanism for accessing altered states of consciousness or higher-dimensional realms.

The effects are achieved by influencing the subtle energy fields that exist within and around the human body. By affecting these fields at a subatomic level, the device allows for potential recalibration of the individual's energetic system to achieve a "tuning" effect, much like how a radio adjusts to receive a specific station. When the frequency applied through the transresonator is finely tuned, the user can theoretically experience both physiological and metaphysical shifts. In short, this device will re-magnetize the crystals in your pineal gland and bring you in tune to which ever frequency is inducted through it.

Abstract:

This section provides the investigation of the relationship between resonant frequencies and electromagnetic field (EMF) enhancement, demonstrating that specific frequencies can strengthen EM fields. The study introduces the use of a transresonator device, drawing from the theoretical foundation of the 3/6/9 resonance matrix, solfeggio frequencies, and universal resonance scales. Key experimental results provide tangible proof of frequency's influence on EM fields, establishing potential applications in advanced energy manipulation and metaphysical practices.

The hypothesis explored in this investigation is that frequency, particularly resonant frequencies, can significantly enhance or strengthen electromagnetic fields. This work is grounded in theoretical studies of resonance, including the 3/6/9 matrix, solfeggio frequencies, and the emerging field of transresonator technology.

The goal of this research is to offer a tangible understanding of how specific frequencies impact the strength of an EM field and provide a stepping stone towards using resonance as a tool for manipulating Biomagnetic fields. The concept originated from the discovery of resonance matrices within planetary frequencies, solfeggio scales, and the idea that certain frequencies could serve as "error-correcting codes" in the universe, restoring balance to energy fields.

Literature Review:
The study of electromagnetic fields and frequency resonance has a rich history, ranging from traditional physics to metaphysical applications. Key references include:

Solfeggio Frequencies: A set of frequencies believed to have healing properties, resonating with the body's chakras and energy fields. Their alignment with the 3/6/9 matrix forms a critical aspect of the theory presented here.
3/6/9 Matrix Theory: Inspired by Nikola Tesla's statement that the numbers 3, 6, and 9 hold the key to the universe, this theory proposes that these numbers represent a universal resonance matrix, guiding everything from planetary orbits to quantum fields.
Universal Resonance Frequencies:
The idea that the universe operates within a set of harmonic resonance frequencies, tied to ancient scales and energetic principles, provides the theoretical background for the hypothesis.
Frequency-EMF Interaction:
Recent studies in electromagnetic theory suggest that EM fields can be manipulated by applying resonant frequencies.
This research explores how this interaction occurs and whether it can be harnessed in practical applications like the transresonator.

Experimental Setup:

The primary experimental setup involves a custom-built transresonator device incorporating a magnet stack, signal generator, and EMF meter. The key components of the experimental process included:

1. Magnet Stack: A magnet assembly providing a base EMF of 4 uT (microteslas).

2. 3-in-1 Device: Oscilloscope, Multi-Meter, and a signal generator used to apply a range of resonant frequencies to the electromagnet.

3. EMF Reader: Used to measure the baseline electromagnetic field strength and detect any changes when specific frequencies are applied.

4. Electromagnetic Coil: Disconnected from its original speaker system, this coil was tested by applying frequencies generated by the 3-in-1 device, with and without direct power.

Procedure:

1. Baseline Measurement: The base EMF of the magnet stack was measured without any frequency or power applied. Initial readings showed an average of 4 uT, with peaks up to 18.64 uT.

2. Application of Resonant Frequencies: Frequencies ranging from 100Hz to144,000 Hz, Solfeggio Frequencies and their harmonic numbers from the 3/6/9 matrix were applied to the electromagnet. The EMF strength was measured before and after each frequency was applied, showing consistent enhancement of the EM field, peaking at .81 uT when 144,000 Hz was applied, up from a baseline of .42 uT.

3. Data Collection: The EMF meter was used to record the magnetic field's strength at different frequency levels. The data was analyzed to determine how the field strength correlated with each applied frequency.

Testing with Higher Frequencies:

An additional phase of testing incorporated a DDS signal generator, capable of outputting frequencies up to 350 MHz. This allowed for exploration of higher harmonic resonance, particularly around the frequencies of 96,000 Hz and 192,000 Hz, which were identified as significant based on prior experiments and theoretical models.

4. Results:

The results confirmed the initial hypothesis that frequency enhances the strength of an EM field.

Specifically:

The magnet stack's baseline reading of 4 uT was consistently increased when resonant frequencies were applied.

The most significant enhancement occurred at the 144,000 Hz frequency, which demonstrated a near doubling of the field strength to .81 uT.

Further testing of higher frequencies (96,000 Hz, 192,000 Hz) also showed an increase in EMF strength, suggesting that certain harmonic frequencies within the 3/6/9 matrix scale have the potential to amplify the field.

These results lend credibility to the theory that resonant frequencies act as amplifiers or enhancers of EM fields and could be used for further energy manipulation.

Discussion:

Frequency as an Energy Amplifier:

The experiments demonstrate that certain frequencies, particularly those derived from the 3/6/9 matrix and solfeggio scale, can enhance electromagnetic fields. This provides a direct link between frequency and field strength, which has significant implications for energy manipulation technologies, including the transresonator device.

Application to Transresonator Technology:
The results suggest that transresonator devices, which combine electromagnetic fields and resonant frequencies, could be used for advanced energy manipulation.

Potential applications include:

Energy Healing: Enhanced EM fields could be used to balance or restore Biomagnetic fields in the human body, promoting healing at a cellular level.

Dimensional Exploration: By tuning the transresonator to specific universal resonance frequencies, the device could act as a gateway to higher dimensions or states of consciousness.

Field Manipulation in Quantum Applications:

The ability to amplify electromagnetic fields through resonance opens possibilities for quantum field manipulation, anti-gravity research, and advanced communication technologies.

Cosmic Resonance and Error-Correcting Codes:

The idea that frequencies act as "error-correcting codes" for the universe ties into the results of this research. The enhancement of EM fields through frequency could serve as a mechanism for correcting or restoring balance to energetic systems, both in the human body and the universe at large.

Conclusion:

The findings of this study confirm the hypothesis that resonant frequencies can enhance electromagnetic fields. The 144,000 Hz frequency and other harmonics within the 3/6/9 matrix proved especially significant, demonstrating the potential for further exploration and application of this principle in various fields of research. Future work will focus on expanding the experimental framework, including the use of higher frequencies and more advanced transresonator prototypes. Additionally, the connection between cosmic resonance frequencies and universal error- correcting codes offers an exciting avenue for deeper metaphysical exploration.

References:

Solfeggio Frequencies and Healing

Tesla's 3/6/9 Theory

Electromagnetic Field Manipulation in Quantum Physics

Theoretical Models on Universal Resonance Matrices

The Searl Effect by John Searl

<u>Pictures, Commentary & Definitions</u>
Transresonator Version 1
LFS-190 Bluetooth Speaker with 9v 50n Electromagnet installed into the diaphragm. This model was able to produce transmagnetic resonance fields up to 22,000hz.
20,448hz is a Harmonic Double of 639hz and provided the most intensive physiological affect. Was able to enter a meditative state more quickly and consciously or so it had seemed.
DO NOT OVER EXPOSE!
MAX 15MIN / 24HRS

Prolonged Exposure Symptoms may include but are not limited to: Headache, migraine, nausea, hot and cold flashes, aches in joints and muscle tissue, and ringing in the ears.

Symptoms are those of Energetic Shift or Frequency Overload. Your body will need time to adapt to the new resonance. You are changing your resonance at a subatomic level, please do not forget that.

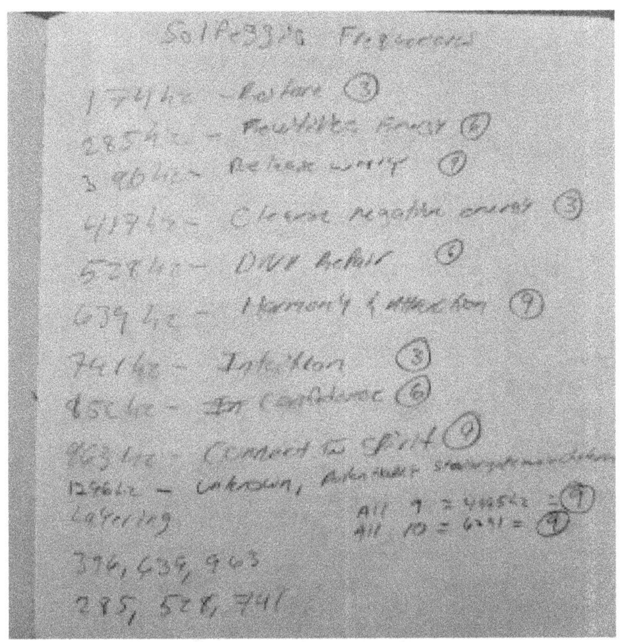

Solfeggio Connection to the 369 Matrix Code
I believe 1296Hz maybe an unused or missing frequency but in any case, the frequencies are stepping stones. It's your resonance that needs to be the same or compliment what you're trying to perceive. Raising your vibration only is like turning the volume up on static. You need to also resonate at said frequency, your Biomagnetic field needs to resonate at the frequency to be able to step through the veil so to speak.

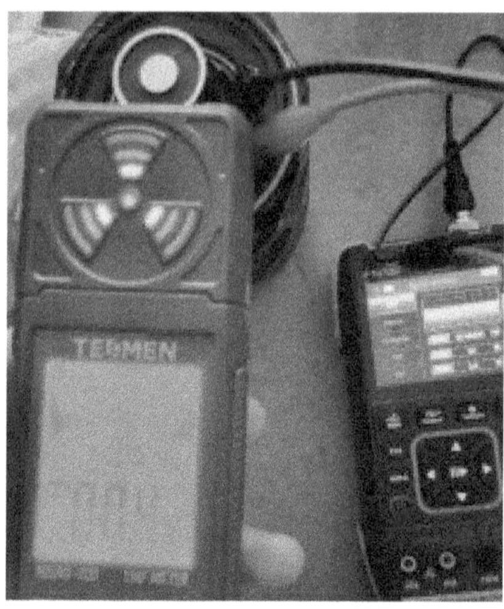

EMF Reading .42, Signal and Speaker off

EMF Reading .81 Signal ON Speaker OFF

Transmagnetic Resonance Field: The type of EM field that is produced when a harmonic frequency is induced through an electromagnet that is within a stable magnetic field.

Harmonic Frequency Induction: The process to which a Transmagnetic Resonance field can be achieved.

Vibration: Is general movement of energy, can be high or low (polarity)

Resonance: When two or more systems vibrate at the same or complementary frequency allowing for harmonious cohesion. Strengthening and amplifying efficiency and output.

Universe Resonance Matrix:

Is a net of interconnected frequencies that defines how reality operates. Once everything is together it creates the fractal holographic matrix that is the method of creation for our universe. Below I will put what I believe is to be the Resonance Scales that make up the Universe Resonance Matrix. Providing the start to the operator's manual to the universe. Such frequencies include: Cosmic, Earth/Nature, Human, and Quantum. With these frequencies one could travel or communicate with higher dimensions, influence matter & energy, and unlock higher state of consciousness.

1/8 Scale: Represents creation and transcendence cycles. The Eternal Energy Loop

2/4 Scale: Mechanical process of Manifestation and how duality forms stable structures within the universe.

5/7 Scale: Growth, Learning, and Evolution

3/6/9 Scale: The key to understanding harmony and universal laws that govern everything. The overarching principle that connects all the layers.

1/8/9 Scale: Creation, Expansion, Completion

1/9 Scale: The Blueprints to creation and completion

9/1 Scale: Return Home, Restart, the back door to the program.

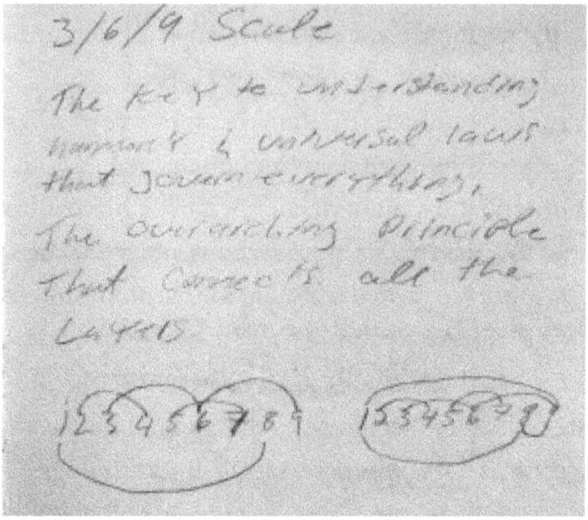

Transresonator & the Resonance Matrix:
The Transresonator is able to generate and manipulate frequencies in such a way that it can tap into or align with the resonance matrix, allowing for the influence or perception or interaction with the universe at deeper levels.
Spinor-Tensor Interactions and Magnetic Field Amplification
The key mechanism behind the transresonator's ability to amplify and strengthen magnetic fields lies in its interaction with spinors and tensors within the atomic structure. This section explains how these quantum mechanical principles work in tandem with the transresonator's frequency emissions to induce profound physical and metaphysical effects.
Spinors and Their Role in Quantum Systems
In quantum mechanics, spinors describe the state of particles that exist in a rotational or oscillatory manner, often influenced by external forces such as electromagnetic fields or frequencies. These spinors represent a fundamental aspect of the atom's quantum state, essentially governing how subatomic particles (like electrons) rotate and oscillate around the nucleus.

By introducing a carefully selected frequency, the transresonator interacts directly with these spinors, aligning their oscillations with the induced frequency. This alignment effectively "tunes" the atomic structure to a new harmonic resonance, allowing for more coherent and efficient behavior of the subatomic particles. As a result, the energy flow between particles increases, causing a significant amplification of the surrounding electromagnetic fields.

Tensors: The Physical Framework

While spinors represent the quantum states, tensors describe the physical forces and interactions that govern energy and matter in space-time. Tensors provide the mathematical framework for physical properties such as electromagnetism, gravity, and pressure.

When the transresonator harmonizes the spinor rotations, it also influences the tensor field interactions. The resonating frequency alters the structure and coherence of these fields, enabling them to interact with the enhanced spinor states more effectively. Essentially, by changing the frequency at which the system operates, the transresonator reshapes the electromagnetic field around it, amplifying the field's strength and coherence.

Spinor-Tensor Resonance and Energy Pathways

The combined effect of the spinor realignment and tensor field modification creates a new energy pathway, or channel of resonance, within the atomic and subatomic structures. This new channel allows energy to flow more freely through the system. The increased coherence and resonance between the spinors and tensors acts as a catalyst for enhanced energy transfer and amplification.

At the atomic level, this results in the strengthening of magnetic fields and increased efficiency of energy transfer. This process is further amplified when multiple frequencies are introduced, creating a harmonic expansion that impacts not just individual atoms but entire systems of particles. The oscillating magnetic field becomes stronger due to the additional energy pathways opened by the transresonator's frequencies.

Implications for Electromagnetic Field Amplification

The practical result of this interaction is the significant amplification of electromagnetic fields generated by the transresonator. The introduction of frequencies such as 144 kHz or 96 kHz interacts with the spinor and tensor systems of the materials involved, particularly in magnetic systems. This interaction strengthens the base electromagnetic field and creates a resonating frequency that multiplies the effect across the system.

In experiments using the transresonator, it was observed that applying specific frequencies to electromagnets significantly increased the strength of their electromagnetic fields. For example, a test conducted with a base field of 0.42 μT showed an increase to 0.81 μT when a 144 kHz frequency was applied. Similarly, in tests with magnetic stacks, the base field of 18.64 μT was significantly enhanced under frequency-driven conditions.

The explanation for this amplification lies in the interaction between the frequency oscillations, spinor states, and tensor fields. The frequency effectively enhances the natural oscillations of the spinors, which in turn modifies the tensor framework of the field, producing a stronger and more coherent electromagnetic effect.

The effect on the Spinors and Tensors prove how and why this device works and is capable of many applications within the quantum field.

www.ingramcontent.com/pod-product-compliance
Lightning Source LLC
Chambersburg PA
CBHW071510130726
47997CB00006B/2481